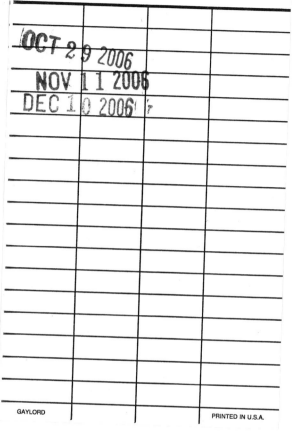

LET'S-READ-AND-FIND-OUT SCIENCE®

STAGE 1

WHERE ARE THE
Night Animals?

MARY ANN FRASER

HarperCollins*Publishers*

*Special thanks to Robin Dalton at the Queens Wildlife Center
for his time and expert review.*

The *Let's-Read-and-Find-Out Science* book series was originated by Dr. Franklyn M. Branley, Astronomer Emeritus and former Chairman of the American Museum–Hayden Planetarium, and was formerly co-edited by him and Dr. Roma Gans, Professor Emeritus of Childhood Education, Teachers College, Columbia University. Text and illustrations for each of the books in the series are checked for accuracy by an expert in the relevant field. For more information about Let's-Read-and-Find-Out Science books, write to HarperCollins Children's Books, 10 East 53rd Street, New York, NY 10022.

HarperCollins®, 🐻®, and Let's Read-and-Find-Out Science® are trademarks of HarperCollins Publishers Inc.

Library of Congress Cataloging-in-Publication Data
Fraser, Mary Ann.
   Where are the night animals? / by Mary Ann Fraser.
      p.   cm. — (Let's-read-and-find-out science. Stage 1)
   Summary: Describes various nocturnal animals and their nighttime activities, including the opossum, brown bat, and tree frog.
   ISBN 0-06-027717-3. — ISBN 0-06-027718-1 (lib. bdg.). — ISBN 0-06-445176-3 (pbk.)
   1. Nocturnal animals—Juvenile literature.   [1. Nocturnal animals.]   I. Title.   II. Series.
QL755.5.F735   1999   97-34683
591.5'18—dc21   CIP
                                                                                            AC

3   4   5   6   7   8   9   10
❖

To Mike

The summer moon rises over the hill.
A lone coyote howls. It sniffs the air.
Then it begins its nightly hunting route.
With the coyote gone, a skunk scoots out
from its den in a log. Another skunk appears.
The two romp, squeaking and squealing.

5

They don't see the harvest mice
scampering among the fallen branches.
But a watchful barn owl does.
It hoots and goes back to eating
a gopher.

The sound of the owl startles an opossum munching berries. She and her babies duck into the underbrush. Then cautiously she waddles to the pond for a drink.

In the muddy water a raccoon feels around for crayfish and snails. Above, a male tree frog calls to a female tree frog, "*Kreck-ek, kreck-ek.*"

A shadow passes over the frog. A little brown bat dips
and dives to snatch moths and mosquitoes from the air.
By dawn it has eaten one quarter of its weight in insects.
The sun rises. The animals retreat to their homes.
The day belongs to others.

Some animals are more active during the day. They are called *diurnal.*

Animals that are more active at night are called *nocturnal.* They have adapted to life in the dark. We never see most of these animals. They are hiding during the day when we are awake.

With few people around after sunset,
the coyote feels safe to crawl out of its den.
It yips and howls. From far off, another
coyote answers.

A skunk peeps out from its den and sees
a beetle. It bounds toward the insect.
The skunk's black-and-white fur
blends in with the dark night.

Many night creatures are black or gray.
These colors make it hard for their enemies
to see them.

The coyote does not see the skunk.
But its sensitive nose smells it. The coyote
moves too close, and the skunk sprays it
in the face with a foul-smelling, oily fluid.
With a yipe the coyote runs away.

The harvest mice in the grass ignore the sharp odor. They scurry about looking for seeds. Their whiskers and fur help them feel their way in the dark. Their shrill squeaks to each other are hard for humans to hear. But not for the barn owl, who is wide awake and listening.

The owl swoops down from its perch. Like all barn owls, one of its ears is lower than the other. This helps the owl find the source of the squeaks. Its extra-large eyes guide its attack. The owl snatches up a mouse in its sharp claws. Then it lands on some grapevines that ramble over a fence.

The opossum family is feasting on the grapes.
Opossums cannot run quickly to escape their
enemies, such as coyotes. They must look for
food while under the cover of darkness.

An anxious baby opossum sees the owl
and tumbles from its mother's back.
It lands beside the pond
unhurt and climbs
back up to safety.

21

The raccoon comes to the pond every night.
Many nocturnal animals are creatures of habit.
Visiting the same spots each night makes it
easier for them to travel in the dark.

The raccoon snatches a crayfish from the pond.
Then it dashes off with its meal through
some reeds.

The tree frog leaps out of the reeds and lands on an oak tree. This small frog is an amphibian. Amphibians breathe through their lungs and skin. If they were active during the day, the hot sun would dry out their skin, and they would die. Night air is cooler and moister.

The little brown bat darts in and out of the oak's twisted branches. *"Click, click, click."* The bat is making noises to help it navigate and to find its food. The sounds bounce off objects, causing echoes. The bat can tell from the echoes how far away the object is.

This is called *echolocation*.

Some scientists think that bats became nocturnal to protect themselves from daytime animals. Other scientists believe that bats hunt at night so that they do not have to compete with birds for food in the daytime.

27

The moon fades from sight as the sun rises. The creatures of the night begin to seek out their dens and burrows. The animals of the day stir from their sleep. Sunrise and sunset are nature's busiest hours.

A young raccoon passes a window. A child comes to the table for breakfast. The two meet. Then each goes its own way. The night shift ends. The day shift begins.

Where do these nocturnal animals go during the day?

Why don't we see nocturnal animals during the day? Where are they hiding? Like people, animals that are active at night don't like to be disturbed when they are sleeping. They look for places that are dark, cool, and safe from their enemies. Some of their hideouts may surprise you.

Coyotes

In the spring coyotes will live in underground dens while raising their pups. At other times they may sleep aboveground, but their homes are always within a few miles of water.

Harvest Mice

These mice live among low-growing plants. They weave amazing ball-like nests out of coarse grasses. The entrance is usually a little round hole on the bottom.

Skunks

Skunks can learn to live almost anywhere. During the day they often hide in hollow logs or other animals' abandoned dens. Around towns they may find a corner under a porch or building in which to rest.

Barn Owls

Despite the name, barn owls also sleep in other types of high places, such as hollow trees or the rafters of abandoned buildings.

Opossums

Opossums will sleep anywhere that is safe from enemies. They shelter in old dens, hollow trees, culverts, and brush piles, and beneath buildings.

Raccoons

Away from the city, raccoons usually spend the day in trees. They will gnaw on the inside of the tree and use the chips for bedding. Sometimes they will sleep in crevices among tree roots, in woodchuck burrows, or simply stretched out on a tree limb. In town raccoons will sleep in drainpipes, sheds, or attics.

Tree Frogs

To keep cool and safe from predators, tree frogs seek shelter in water and wet vegetation. If you look carefully, you might see one crouched under a stone, wedged into a crevice, or tucked into a clump of grass. Near homes they will hide in a drain, near a man-made pond, or even in a well-watered flowerpot.

Bats

Bats like cool, dark places, such as caves, and crevices in trees, rocks, or buildings. They usually sleep in groups and hang by their hind feet.

Where Are the Night Animals?
Find Out More

Play hide-and-seek in the dark.

1. When it's dark outside, turn out the lights in a room in your house, and play hide-and-seek with your friends or family.

2. When it is your turn to hide, be very quiet. Choose the darkest place you can find. You are acting like an opossum or a mouse.

3. When it is your turn to seek, listen for sounds like breathing or shuffling that people make when they hide. Feel around in the shadows. You are acting like an owl or a coyote hunting for food.

4. When you move about, go around the furniture the same way you do when the lights are on. You are acting like a raccoon or a coyote who travels the same route every night.

Look for night animals.

If you live outside a city, you probably live very close to nocturnal animals without ever seeing them. How do you know that they're there?

1. When the ground is soft, go out with a parent in the woods or fields to look for opossum or raccoon tracks.

2. In the evening, go outside and listen. Do you hear owls hooting? If you live near a pond, do you hear frogs calling?

3. Sniff the air. Sometimes you can smell a skunk nearby.

4. Look in the sky. You may see a bat fly overhead.

If you live in a city and you want to see nocturnal animals, visit the zoo closest to your home. They have special rooms where you can see these animals during the day.

Help a mouse build a nest.

A mouse is often in danger at night. Lots of animals want to eat it. It is small and has few defenses. Its nest is a place to hide. By making a mouse house, you can help protect a mouse.

1. You will need five pieces of wood to make a mouse house. Four of the pieces should be 5 inches wide and 6 inches high. One piece should be 5 inches square. Get an adult to help you cut the wood to the right size.

2. Have an adult make a 2-inch-square hole in the middle of one of the 5" x 6" pieces.

3. The four 5" x 6" pieces will form the walls for your mouse house. Glue the 6-inch sides together so that your house is 6 inches high.

4. Glue the 5" x 5" piece on the top to make a roof.

5. Take your mouse house outside and put it near the woods or in a field. Check it periodically to see if a mouse has come to build its nest there.